ISBN 978-1-333-75849-3
PIBN 10544223

# 1 MONTH OF
# FREE
# READING

at

## www.ForgottenBooks.com

By purchasing this book you are eligible for one month membership to ForgottenBooks.com, giving you unlimited access to our entire collection of over 700,000 titles via our web site and mobile apps.

To claim your free month visit:

www.forgottenbooks.com/free544223

# Length Relations of Some Marine Fishes
# From Coastal Georgia

**SPECIAL SCIENTIFIC REPORT—FISHERIES No. 575**

UNITED STATES DEPARTMENT OF THE INTERIOR

U.S. FISH AND WILDLIFE SERVICE

BUREAU OF COMMERCIAL FISHERIES

# Length Relations of Some Marine Fishes
# From Coastal Georgia

By

SHERRELL C. JORGENSON and GRANT L. MILLER

Contribution No. 93, Bureau of Commercial Fisheries
Biological Laboratory, Brunswick, Ga. 31520

United States Fish and Wildlife Service

Special Scientific Report--Fisheries No. 575

Washington, D.C.

November 1968

## CONTENTS

# Length Relations of Some Marine Fishes From Coastal Georgia

By

SHERRELL C. JORGENSON AND GRANT L. MILLER, Fishery Biologists

Bureau of Commercial Fisheries Biological Laboratory
Brunswick, Georgia 31520

## ABSTRACT

Information is given for converting standard length into fork and total lengths, fork length into standard and total lengths, and total length into standard and fork lengths of 82 species of marine fishes collected in coastal Georgia during 1953-61.

## INTRODUCTION

The Bureau of Commercial Fisheries Biological Laboratory, Brunswick, Ga., is charged with a study of the life histories of fishes of the western North Atlantic Ocean with emphasis on those off the southeastern coast of the United States. One of the laboratory's programs, a study of coastal and estuarine ecology, includes a major project on the occurrence, abundance, seasonal distribution, apparent hydrographic preferences, and early life history of fishes of coastal Georgia.

During 1953-61, young of marine fishes were collected by seining at selected localities in three types of environment in McIntosh and Glynn counties, Georgia: the upper tidewaters of the Altamaha River, the ocean beach, and the saltmarshes. The collections included larvae, juveniles, and adults of some species, but only juveniles of others.

Published information on growth and changes in body form of fishes during development generally records length as standard, fork, or total. Comparison of these data is difficult or impossible without a means of converting one measurement to another. We wanted to be able to make such comparisons in detailed studies we were planning; so we determined the relations of standard, fork, and total length by the method of least squares for those species for which we had adequate data. This report presents the statistics describing these relations for 82 marine species and gives factors for converting one length measurement to another. Length relations for fresh-water species

from the Altamaha River were not determined.

## METHODS

To determine the relation of various length measurements, we measured representative samples of all sizes of all species for standard and total lengths and (for species with forked caudal fins) fork length. Sample sizes ranged from a few specimens for some species to thousands for others.

Standard length was measured from the tip of the snout to the end of the hypural bones (the caudal base), fork length from the tip of the snout to the tip of the shortest median caudal ray, and total length from the tip of the snout to a vertical at the tip of the longest lobe, or ray, of the caudal fin. Measurements (to the nearest millimeter) were made with dial calipers or a measuring board.

We measured a size series of specimens for each seine sample of each species. Some millimeter size groups were represented by one specimen; others by many specimens. We subsampled our data to treat the species uniformly. For each millimeter size group of standard length represented in our data, we selected at random one specimen and recorded its fork and total lengths; for each millimeter size group of fork length, we selected one specimen and recorded its standard and total lengths; and for each millimeter size group of total length, we selected one specimen and recorded its standard and fork lengths. We had three sets of data for species with forked caudal fins and two sets of data for those without forked caudals. We further refined the data by selecting only those portions of the size series in which the observations were uniformly distributed, by discarding scattered observations at either end of the size range, and by omitting those species for which fewer than five size groups were available.

Linear regressions were calculated for standard, fork, and total length relations by the method of least squares. The Biometrics Unit, Bureau of Commercial Fisheries Biological Laboratory, Seattle, Wash., provided computer services to calculate regressions for species with a sample size of 25 or more,

and we calculated the regressions for the rest of the species.

## LENGTH RELATIONS

Three tables show the data essential to this paper. Table 1 gives the statistics describing the relations of standard, fork, and total lengths for 82 species of marine fishes from coastal Georgia. Table 2 gives the factors for converting one length to another, for the size range represented in our samples. Table 3 lists scientific and common names for all species mentioned in this report.

Table 1.--Relation of standard, fork, and total lengths in marine fishes from coastal Georgia

[SL = standard length, FL = fork length, TL = total length, $\bar{x}$ = mean of values of X, $\bar{y}$ = mean of values of Y, N = number of millimeter size groups, b = change in Y for unit change in X, a = Y-intercept of regression line, and Sy.x = standard deviation from regression (standard error of estimate)]

| Species | Independent variable X | Dependent variable Y | Size range, independent variable X | $\bar{x}$ | $\bar{y}$ | N | b | a | Sy.x |
|---|---|---|---|---|---|---|---|---|---|
| | | | Mm. | Mm. | Mm. | | | | |
| **ANGUILLIDAE** | | | | | | | | | |
| Anguilla rostrata | SL | TL | 45- 62 | 51.83 | 52.91 | 12 | 0.988 | 1.676 | 0.296 |
| | TL | SL | 46- 63 | 52.91 | 51.83 | 12 | 1.008 | -1.498 | .299 |
| **ARIIDAE** | | | | | | | | | |
| Galeichthys felis | SL | FL | 39- 93 | 64.97 | 70.72 | 36 | 1.063 | 1.630 | .872 |
| | SL | TL | 39- 93 | 64.97 | 84.53 | 36 | 1.320 | -1.245 | 2.818 |
| | FL | SL | 44- 99 | 69.53 | 63.91 | 34 | .930 | -.741 | 1.231 |
| | FL | TL | 44- 99 | 69.53 | 82.91 | 34 | 1.228 | -2.466 | 2.171 |
| | TL | SL | 51-114 | 81.31 | 62.82 | 39 | .781 | -.645 | .902 |
| | TL | FL | 51-114 | 81.31 | 68.46 | 39 | .833 | .756 | .648 |
| **ATHERINIDAE** | | | | | | | | | |
| Membras martinica | SL | FL | 13- 84 | 48.50 | 55.06 | 72 | 1.135 | .029 | .961 |
| | SL | TL | 13- 84 | 48.50 | 58.32 | 72 | 1.199 | .154 | 1.339 |
| | FL | SL | 15- 92 | 53.50 | 47.13 | 78 | .874 | .352 | .971 |
| | FL | TL | 15- 92 | 53.50 | 56.73 | 78 | 1.061 | -.033 | .965 |
| | TL | SL | 16- 93 | 54.50 | 45.29 | 78 | .832 | -.078 | .944 |
| | TL | FL | 16- 93 | 54.50 | 51.41 | 78 | .948 | -.230 | .602 |
| Menidia menidia | SL | FL | 12- 95 | 53.01 | 60.86 | 83 | 1.127 | 1.112 | .990 |
| | SL | TL | 12- 95 | 53.01 | 64.96 | 83 | 1.196 | 1.534 | 1.210 |
| | FL | SL | 13-107 | 59.02 | 51.42 | 93 | .875 | -.247 | .940 |
| | FL | TL | 13-107 | 59.02 | 63.01 | 93 | 1.063 | .286 | .815 |
| | TL | SL | 14-113 | 63.50 | 51.78 | 100 | .827 | -.708 | 1.041 |
| | TL | FL | 14-113 | 63.50 | 59.46 | 100 | .934 | .122 | .926 |
| **BATRACHOIDIDAE** | | | | | | | | | |
| Opsanus tau | SL | TL | 59-153 | 111.10 | 131.40 | 10 | 1.180 | .280 | 1.337 |
| | TL | SL | 71-183 | 131.40 | 111.10 | 10 | .846 | -.051 | 1.132 |
| **BELONIDAE** | | | | | | | | | |
| Strongylura marina | SL | TL | 21-179 | 123.41 | 133.82 | 39 | 1.061 | 2.878 | 2.678 |
| | TL | SL | 23-193 | 132.26 | 121.87 | 39 | .934 | -1.676 | 2.401 |
| **BLENNIIDAE** | | | | | | | | | |
| Chasmodes bosquianus | SL | TL | 11- 71 | 43.55 | 52.55 | 9 | 1.180 | 1.183 | 2.046 |
| | TL | SL | 14- 84 | 52.55 | 43.55 | 9 | .841 | -.666 | 1.728 |
| Hypsoblennius hentzi | SL | TL | 8- 65 | 37.05 | 45.44 | 18 | 1.227 | -.017 | .750 |
| | TL | SL | 9- 78 | 45.44 | 37.05 | 18 | .814 | .057 | .611 |

Table 1.--Relation of standard, fork, and total lengths in marine fishes from coastal Georgia--Continued

| Species | Inde- pendent vari- able X | Depend- ent variable Y | Size range, independ- ent variable X | $\bar{x}$ | $\bar{y}$ | N | b | a | Sy.x |
|---|---|---|---|---|---|---|---|---|---|
| | | | Mm. | Mm. | Mm. | | | | |
| BOTHIDAE | | | | | | | | | |
| Etropus crossotus | SL | TL | 48- 71 | 63.77 | 80.44 | 9 | 1.240 | 1.346 | .827 |
| | TL | SL | 60- 88 | 80.44 | 63.77 | 9 | .799 | -.534 | .664 |
| Paralichthys dentatus | SL | TL | 29- 56 | 39.18 | 47.81 | 11 | 1.271 | -1.992 | 1.577 |
| | TL | SL | 32- 69 | 47.81 | 39.18 | 11 | .777 | 2.013 | 1.233 |
| Paralichthys lethostigma | SL | TL | 20- 93 | 45.48 | 57.15 | 33 | 1.257 | -.005 | 2.469 |
| | TL | SL | 26-117 | 54.84 | 43.30 | 37 | .800 | -.563 | 2.134 |
| Paralichthys squamilentus | SL | TL | 20- 42 | 32.70 | 41.50 | 10 | 1.223 | 1.501 | .805 |
| | TL | SL | 26- 53 | 41.50 | 32.70 | 10 | .812 | -.986 | .656 |
| Scophthalmus aquosus | SL | TL | 47- 78 | 65.10 | 83.10 | 10 | 1.212 | 4.225 | 1.526 |
| | TL | SL | 61- 99 | 83.10 | 65.10 | 10 | .814 | -2.502 | 1.251 |
| CARANGIDAE | | | | | | | | | |
| Caranx hippos | SL | FL | 21-106 | 52.00 | 56.43 | 35 | 1.069 | .838 | .538 |
| | SL | TL | 21-106 | 52.00 | 64.63 | 35 | 1.247 | -.195 | 1.699 |
| | FL | SL | 23-114 | 57.33 | 52.85 | 33 | .935 | -.756 | .456 |
| | FL | TL | 23-114 | 57.33 | 65.58 | 33 | 1.166 | -1.246 | 1.815 |
| | TL | SL | 26-132 | 63.61 | 51.25 | 36 | .800 | .394 | 1.351 |
| | TL | FL | 26-132 | 63.61 | 55.92 | 36 | .857 | 1.404 | 2.291 |
| Chloroscombrus chrysurus | SL | FL | 15-113 | 56.96 | 63.35 | 78 | 1.087 | 1.417 | 1.017 |
| | SL | TL | 15-113 | 56.96 | 73.55 | 78 | 1.316 | -1.408 | 1.556 |
| | FL | SL | 17-131 | 64.47 | 58.02 | 88 | .921 | -1.352 | .909 |
| | FL | TL | 17-131 | 64.47 | 75.16 | 88 | 1.216 | -3.204 | 1.156 |
| | TL | SL | 18-127 | 70.19 | 54.30 | 99 | .754 | 1.412 | 1.005 |
| | TL | FL | 18-127 | 70.19 | 60.45 | 99 | .822 | 2.764 | .966 |
| Oligoplites saurus | SL | FL | 9-106 | 51.17 | 55.76 | 66 | 1.060 | 1.544 | 0.779 |
| | SL | TL | 9-106 | 51.17 | 61.56 | 66 | 1.181 | 1.143 | 1.030 |
| | FL | SL | 10-113 | 54.20 | 49.75 | 71 | .942 | -1.324 | .731 |
| | FL | TL | 10-113 | 54.20 | 59.86 | 71 | 1.116 | -.644 | .621 |
| | TL | SL | 11-127 | 60.36 | 50.32 | 75 | .846 | -.767 | .878 |
| | TL | FL | 11-127 | 60.36 | 54.69 | 75 | .896 | .607 | .622 |
| Selene vomer | SL | FL | 24- 78 | 40.22 | 44.11 | 9 | 1.099 | -.100 | .663 |
| | SL | TL | 24- 78 | 40.22 | 53.55 | 9 | 1.338 | -.260 | .883 |
| | FL | SL | 27- 86 | 44.11 | 40.22 | 9 | .909 | .133 | .603 |
| | FL | TL | 27- 86 | 44.11 | 53.55 | 9 | 1.216 | -.070 | 1.381 |
| | TL | SL | 32-105 | 53.55 | 40.22 | 9 | .746 | .245 | .660 |
| | TL | FL | 32-105 | 53.55 | 44.11 | 9 | .820 | .178 | 1.134 |
| Trachinotus carolinus | SL | FL | 11-102 | 50.63 | 57.49 | 78 | 1.100 | 1.816 | .972 |
| | SL | TL | 11-102 | 50.63 | 66.74 | 78 | 1.340 | -1.089 | 1.674 |
| | FL | SL | 13-100 | 52.56 | 46.00 | 78 | .903 | -1.461 | 1.015 |
| | FL | TL | 13-100 | 52.56 | 60.69 | 78 | 1.214 | -3.146 | 1.080 |
| | TL | SL | 14-102 | 56.24 | 42.66 | 85 | .740 | 1.064 | .916 |
| | TL | FL | 14-102 | 56.24 | 48.87 | 85 | .822 | 2.655 | .829 |
| Trachinotus falcatus . | SL | FL | 12- 59 | 34.78 | 40.02 | 46 | 1.082 | 2.386 | .647 |
| | SL | TL | 12- 59 | 34.78 | 46.28 | 46 | 1.303 | .950 | .662 |
| | FL | SL | 15- 67 | 39.72 | 34.49 | 47 | .922 | -2.148 | .589 |
| | FL | TL | 15- 67 | 39.72 | 45.94 | 47 | 1.214 | -2.292 | .858 |
| | TL | SL | 16- 79 | 44.78 | 33.70 | 54 | .769 | -.712 | .650 |
| | TL | FL | 16- 79 | 44.78 | 38.81 | 54 | .831 | 1.618 | .606 |
| Trachinotus glaucus | SL | FL | 19- 60 | 36.93 | 40.80 | 15 | 1.122 | 1.372 | .593 |
| | SL | TL | 19- 60 | 36.93 | 48.13 | 15 | 1.326 | -.839 | .762 |
| | FL | SL | 22- 68 | 42.80 | 36.93 | 15 | .890 | -1.171 | .528 |
| | FL | TL | 22- 68 | 42.80 | 48.13 | 15 | 1.182 | -2.447 | .551 |
| | TL | SL | 24- 78 | 48.13 | 36.93 | 15 | .753 | .693 | .574 |
| | TL | FL | 24- 78 | 48.13 | 42.80 | 15 | .846 | 2.106 | .466 |

CARANGIDAE--Continued

| Vomer setapinnis | SL | FL | 41- 60 | 52.40 | 58.20 | 5 | 1.169 | -3.061 | .986 |
| | SL | TL | 41- 60 | 52.40 | 68.60 | 5 | 1.410 | -5.289 | 1.496 |
| | FL | SL | 45- 68 | 58.20 | 52.40 | 5 | .850 | 2.930 | .841 |
| | FL | TL | 45- 68 | 58.20 | 68.60 | 5 | 1.208 | -1.682 | .491 |
| | TL | SL | 53- 81 | 68.60 | 52.40 | 5 | .702 | 4.229 | 1.056 |
| | TL | FL | 53- 81 | 68.60 | 58.20 | 5 | .827 | 1.461 | .406 |

CLUPEIDAE

| Alosa aestivalis | SL | FL | 28- 58 | 40.00 | 45.00 | 6 | 1.084 | 1.620 | .482 |
| | SL | TL | 28- 58 | 40.00 | 50.16 | 6 | 1.294 | -1.612 | .574 |
| | FL | SL | 32-64 | 45.00 | 40.00 | 6 | .921 | -1.445 | .444 |
| | FL | TL | 32- 64 | 45.00 | 50.16 | 6 | 1.191 | -3.448 | 1.027 |
| | TL | SL | 35- 74 | 50.16 | 40.00 | 6 | .772 | 1.292 | .443 |
| | TL | FL | 35- 74 | 50.16 | 45.00 | 6 | .836 | 3.051 | .860 |

| Alosa sapidissima | SL | FL | 37- 60 | 50.10 | 54.90 | 10 | 1.110 | -.726 | .573 |
| | SL | TL | 37- 60 | 50.10 | 62.00 | 10 | 1.313 | -3.761 | .929 |
| | FL | SL | 41- 66 | 54.90 | 50.10 | 10 | .897 | .838 | .515 |
| | FL | TL | 41- 66 | 54.90 | 62.00 | 10 | 1.179 | -2.738 | 1.007 |
| | TL | SL | 45- 74 | 62.00 | 50.10 | 10 | .756 | 3.197 | .706 |
| | TL | FL | 45- 74 | 62.00 | 54.90 | 10 | .841 | 2.758 | .850 |

| Brevoortia smithi | SL | FL | 21- 74 | 46.53 | 50.84 | 38 | 1.050 | 1.992 | .499 |
| | SL | TL | 21- 74 | 46.53 | 59.61 | 38 | 1.285 | -.191 | .901 |
| | FL | SL | 24- 85 | 49.44 | 46.23 | 39 | .994 | -2.921 | 1.433 |
| | FL | TL | 24- 85 | 49.44 | 59.26 | 39 | 1.283 | -4.188 | 1.954 |
| | TL | SL | 27- 95 | 58.48 | 45.59 | 46 | .776 | .178 | .596 |
| | TL | FL | 27- 95 | 58.48 | 49.83 | 46 | .817 | 2.032 | .656 |

| Brevoortia tyrannus | SL | FL | 14-148 | 69.12 | 76.36 | 105 | 1.103 | .098 | 1.250 |
| | SL | TL | 14-148 | 69.12 | 88.45 | 105 | 1.320 | -2.796 | 2.630 |
| | FL | SL | 16-145 | 76.03 | 68.64 | 117 | .898 | .389 | 1.430 |
| | FL | TL | 16-145 | 76.03 | 88.14 | 117 | 1.195 | -2.712 | 1.588 |
| | TL | SL | 18-174 | 88.56 | 69.06 | 137 | .751 | 2.531 | 1.602 |
| | TL | FL | 18-174 | 88.56 | 76.35 | 137 | .834 | 2.480 | 1.456 |

| Dorosoma cepedianum | SL | FL | 73-117 | 89.14 | 98.71 | 14 | 1.056 | 4.605 | 1.356 |
| | SL | TL | 73-117 | 89.14 | 118.07 | 14 | 1.235 | 8.000 | 1.745 |
| | FL | SL | 81-127 | 98.71 | 89.14 | 14 | .939 | -3.568 | 1.279 |
| | FL | TL | 81-127 | 98.71 | 118.07 | 14 | 1.169 | 2.668 | 1.560 |
| | TL | SL | 97-150 | 118.07 | 89.14 | 14 | .802 | -5.505 | 1.406 |
| | TL | FL | 97-150 | 118.07 | 98.71 | 14 | .848 | -1.461 | 1.328 |

| Dorosoma petenense | SL | FL | 38- 71 | 52.81 | 58.13 | 22 | 1.078 | 1.180 | .660 |
| | SL | TL | 38- 71 | 52.81 | 69.00 | 22 | 1.254 | 2.766 | 1.086 |
| | FL | SL | 42- 78 | 58.13 | 52.81 | 22 | .923 | -.844 | .611 |
| | FL | TL | 42- 78 | 58.13 | 69.00 | 22 | 1.163 | 1.401 | .770 |
| | TL | SL | 50- 93 | 69.00 | 52.81 | 22 | .790 | -1.700 | .862 |
| | TL | FL | 50- 93 | 69.00 | 58.13 | 22 | .856 | -.927 | .660 |

| Harengula pensacolae | SL | FL | 44- 63 | 54.28 | 59.33 | 21 | .999 | 5.083 | .605 |
| | SL | TL | 44- 63 | 54.28 | 68.61 | 21 | 1.280 | -.863 | .582 |
| | FL | SL | 49- 67 | 59.33 | 54.28 | 21 | .988 | -4.356 | .602 |
| | FL | TL | 49- 67 | 59.33 | 68.61 | 21 | 1.270 | -6.727 | .752 |
| | TL | SL | 56- 79 | 68.61 | 54.28 | 21 | .776 | 1.052 | .453 |
| | TL | FL | 56- 79 | 68.61 | 59.33 | 21 | .778 | 5.924 | .588 |

| Species | Independent variable X | Dependent variable Y | Size range, independent variable X | $\bar{x}$ | $\bar{y}$ | N | b | a | Sy.x |
|---|---|---|---|---|---|---|---|---|---|
| | | | Mm. | Mm. | Mm. | | | | |
| **CLUPEIDAE--Continued** | | | | | | | | | |
| Opisthonema oglinum | SL | FL | 28- 71 | 51.87 | 57.33 | 39 | 1.073 | 1.664 | 0.840 |
| | SL | TL | 28- 71 | 51.87 | 65.26 | 39 | 1.245 | .665 | 1.427 |
| | FL | SL | 31- 82 | 58.84 | 52.80 | 45 | .864 | 1.972 | 1.799 |
| | FL | TL | 31- 82 | 58.84 | 66.24 | 45 | 1.158 | -1.875 | 2.055 |
| | TL | SL | 35- 97 | 67.40 | 53.80 | 50 | .735 | 4.265 | 1.505 |
| | TL | FL | 35- 97 | 67.40 | 59.64 | 50 | .832 | 3.544 | 1.522 |
| Sardinella anchovia | SL | | 24- 30 | 27.16 | 30.50 | 6 | 1.099 | .648 | .519 |
| | SL | | 24- 30 | 27.16 | 33.00 | 6 | 1.304 | -2.417 | .300 |
| | FL | | 27- 34 | 30.50 | 27.16 | 6 | .881 | .305 | .464 |
| | FL | | 27- 34 | 30.50 | 33.00 | 6 | 1.164 | -2.505 | .386 |
| | TL | | 29- 37 | 33.00 | 27.16 | 6 | .761 | 2.054 | .229 |
| | TL | FL | 29- 37 | 33.00 | 30.50 | 6 | .848 | 2.523 | .330 |
| **CYNOGLOSSIDAE** | | | | | | | | | |
| Symphurus plagiusa | SL | TL | 11-124 | 47.20 | 51.70 | 40 | 1.096 | -.055 | .990 |
| | TL | SL | 13-138 | 51.70 | 47.20 | 40 | .911 | .096 | .902 |
| **CYPRINODONTIDAE** | | | | | | | | | |
| Cyprinodon variegatus | SL | TL | 16- 41 | 28.08 | 34.88 | 25 | 1.166 | 2.133 | .733 |
| | TL | SL | 21- 51 | 34.88 | 28.08 | 25 | .851 | -1.617 | .626 |
| Fundulus heteroclitus | SL | TL | 5- 79 | 39.68 | 49.05 | 66 | 1.196 | 1.580 | 1.167 |
| | TL | SL | 7- 96 | 47.76 | 38.41 | 78 | .823 | -.899 | 1.015 |
| Fundulus luciae | SL | TL | 10- 31 | 18.95 | 23.30 | 23 | 1.210 | .378 | .687 |
| | TL | SL | 12- 39 | 23.30 | 18.95 | 23 | .820 | -.149 | .566 |
| Fundulus majalis | SL | TL | 8- 87 | 51.18 | 62.32 | 87 | 1.182 | 1.800 | 1.107 |
| | TL | SL | 10-116 | 60.34 | 49.58 | 101 | .844 | -1.323 | .962 |
| **DIODONTIDAE** | | | | | | | | | |
| Chilomycterus schoepfi | SL | TL | 12- 54 | 28.85 | 35.18 | 27 | 1.187 | .944 | .942 |
| | TL | SL | 15- 64 | 35.18 | 28.85 | 27 | .839 | -.673 | .792 |
| **ECHELIDAE** | | | | | | | | | |
| Myrophis punctatus | SL | TL | 132-170 | 151.00 | 152.10 | 10 | 1.013 | -.923 | .284 |
| | TL | SL | 133-172 | 152.10 | 151.00 | 10 | .986 | .984 | .250 |
| **ELOPIDAE** | | | | | | | | | |
| Elops saurus | SL | FL | 23-103 | 59.18 | 63.45 | 22 | 1.056 | .950 | .773 |
| | SL | TL | 23-103 | 59.18 | 74.54 | 22 | 1.294 | -2.068 | 1.395 |
| | FL | SL | 25-109 | 63.45 | 59.18 | 22 | .946 | -.844 | .732 |
| | FL | TL | 25-109 | 63.45 | 74.54 | 22 | 1.225 | -3.192 | 1.476 |
| | TL | SL | 29-131 | 74.54 | 59.18 | 22 | .771 | 1.702 | 1.077 |
| | TL | FL | 29-131 | 74.54 | 63.45 | 22 | .815 | 2.730 | 1.204 |

| Species | Independent variable X | Dependent variable Y | Size range, independent variable X | $\bar{x}$ | $\bar{y}$ | | | | |
|---|---|---|---|---|---|---|---|---|---|
| | | | | Mm. | Mm. | Mm. | | | |
| **ENGRAULIDAE** | | | | | | | | | |
| Anchoa hepsetus | SL | FL | 13- 94 | 53.58 | 58.68 | 78 | 1.078 | .932 | .770 |
| | SL | TL | 13- 94 | 53.58 | 65.42 | 78 | 1.228 | -.395 | 1.208 |
| | FL | SL | 15-104 | 59.50 | 54.40 | 88 | .927 | -.756 | .648 |
| | FL | TL | 15-104 | 59.50 | 66.39 | 88 | 1.141 | -1.508 | .947 |
| | TL | SL | 16-117 | 65.76 | 54.09 | 98 | .817 | .390 | .973 |
| | TL | FL | 16-117 | 65.76 | 59.10 | 98 | .881 | 1.178 | .958 |
| Anchoa lyolepis | SL | FL | 28- 38 | 31.90 | 35.80 | 10 | 1.077 | 1.453 | .898 |
| | SL | TL | 28- 38 | 31.90 | 39.00 | 10 | 1.378 | -4.971 | .697 |
| | FL | SL | 32- 42 | 35.80 | 31.90 | 10 | .866 | .894 | .805 |
| | FL | TL | 32- 42 | 35.80 | 39.00 | 10 | 1.213 | -4.436 | 1.075 |
| | TL | SL | 34- 48 | 39.00 | 31.90 | 10 | .707 | 4.343 | .499 |
| | TL | FL | 34- 48 | 39.00 | 35.80 | 10 | .773 | 5.641 | .858 |
| Anchoa mitchilli | SL | FL | 10- 71 | 40.03 | 43.52 | 61 | 1.078 | .389 | .623 |
| | SL | TL | 10- 71 | 40.03 | 48.30 | 61 | 1.214 | -.315 | .940 |
| | FL | SL | 11- 76 | 42.56 | 39.20 | 64 | .929 | -.350 | .600 |
| | FL | TL | 11- 76 | 42.56 | 47.20 | 64 | 1.126 | -.701 | .748 |
| | TL | SL | 12- 84 | 47.51 | 39.33 | 72 | .812 | .768 | .766 |
| | TL | FL | 12- 84 | 47.51 | 42.75 | 72 | .887 | .586 | .715 |
| **EPHIPPIDAE** | | | | | | | | | |
| Chaetodipterus faber | SL | TL | 5- 32 | 15.15 | 20.36 | 19 | 1.278 | 1.000 | .745 |
| | TL | SL | 7- 43 | 20.36 | 15.15 | 19 | .779 | -.714 | .582 |
| **GERRIDAE** | | | | | | | | | |
| Diapterus olisthostomus | SL | FL | 17- 56 | 28.42 | 32.00 | 7 | 1.115 | 0.312 | 0.744 |
| | SL | TL | 17- 56 | 28.42 | 37.85 | 7 | 1.318 | .398 | 2.023 |
| | FL | SL | 19- 62 | 32.00 | 28.42 | 7 | .895 | -.226 | .667 |
| | FL | TL | 19- 62 | 32.00 | 37.85 | 7 | 1.184 | -.032 | 1.342 |
| | TL | SL | 23- 72 | 37.85 | 28.42 | 7 | .752 | -.039 | 1.528 |
| | TL | FL | 23- 72 | 37.85 | 32.00 | 7 | .841 | .157 | 1.131 |
| Eucinostomus gula | SL | FL | 10- 71 | 37.58 | 42.04 | 53 | 1.106 | .462 | 1.002 |
| | SL | TL | 10- 71 | 37.58 | 49.02 | 53 | 1.328 | -.901 | .958 |
| | FL | SL | 11- 89 | 42.24 | 37.32 | 59 | .880 | .140 | 1.506 |
| | FL | TL | 11- 89 | 42.24 | 48.90 | 59 | 1.172 | -.601 | 1.848 |
| | TL | SL | 12- 93 | 45.85 | 35.13 | 61 | .751 | .681 | .705 |
| | TL | FL | 12- 93 | 45.85 | 39.54 | 61 | .834 | 1.304 | .749 |
| **GOBIESOCIDAE** | | | | | | | | | |
| Gobiesox strumosus | SL | TL | 7- 54 | 28.71 | 35.56 | 34 | 1.240 | -.027 | 1.004 |
| | TL | SL | 9- 68 | 36.24 | 29.32 | 37 | .801 | .301 | .846 |
| **GOBIIDAE** | | | | | | | | | |
| Gobionellus shufeldti | SL | TL | 13- 65 | 34.77 | 45.79 | 39 | 1.352 | -1.218 | 1.321 |
| | TL | SL | 17- 67 | 41.28 | 31.54 | 39 | .740 | 1.004 | .762 |
| Gobiosoma bosci | SL | TL | 8- 45 | 28.31 | 34.88 | 32 | 1.230 | .061 | .608 |
| | TL | SL | 9- 56 | 32.69 | 26.60 | 35 | .806 | .246 | .498 |
| **MONACANTHIDAE** | | | | | | | | | |
| Stephanolepis hispidus | SL | TL | 8- 53 | 25.78 | 32.72 | 36 | 1.247 | .568 | .661 |
| | TL | SL | 10- 50 | 30.87 | 24.18 | 39 | .795 | -.353 | .569 |

| Species | Independent variable X | Dependent variable Y | Size range, independent variable X (Mm.) | $\bar{x}$ (Mm.) | $\bar{y}$ (Mm.) | N | b | a | Sy·x |
|---|---|---|---|---|---|---|---|---|---|
| **MUGILIDAE** | | | | | | | | | |
| Mugil cephalus | SL | FL | 13-230 | 103.75 | 119.88 | 174 | 1.135 | 2.148 | 1.763 |
| | SL | TL | 13-230 | 103.75 | 130.94 | 174 | 1.266 | -.424 | 2.031 |
| | FL | SL | 16-248 | 117.35 | 101.30 | 192 | .877 | -1.584 | 1.732 |
| | FL | TL | 16-248 | 117.35 | 128.07 | 192 | 1.115 | -2.733 | 1.634 |
| | TL | SL | 17-219 | 117.21 | 92.88 | 195 | .787 | .646 | 1.334 |
| | TL | FL | 17-219 | 117.21 | 107.75 | 195 | .900 | 2.233 | 1.090 |
| Mugil curema | SL | FL | 16-125 | 67.66 | 79.38 | 104 | 1.157 | 1.066 | .906 |
| | SL | TL | 16-125 | 67.66 | 86.69 | 104 | 1.288 | -.468 | 1.254 |
| | FL | SL | 19-143 | 79.07 | 67.45 | 121 | .860 | -.542 | .942 |
| | FL | TL | 19-143 | 79.07 | 86.39 | 121 | 1.114 | -1.703 | .861 |
| | TL | SL | 20-155 | 86.10 | 67.29 | 133 | .773 | .751 | .925 |
| | TL | FL | 20-155 | 86.10 | 78.74 | 133 | .897 | 1.488 | .731 |
| **OPHIDIIDAE** | | | | | | | | | |
| Rissola marginata | SL | TL | 134-182 | 157.90 | 162.00 | 10 | .973 | 8.411 | .805 |
| | TL | SL | 138-184 | 162.00 | 157.90 | 10 | 1.026 | -8.231 | .826 |
| **POECILIIDAE** | | | | | | | | | |
| Gambusia affinis | SL | TL | 14- 36 | 22.52 | 29.52 | 23 | 1.276 | .796 | 1.675 |
| | TL | SL | 17- 46 | 29.52 | 22.52 | 23 | .758 | .156 | 1.291 |
| Heterandria formosa | SL | TL | 12- 19 | 15.80 | 20.20 | 5 | .951 | 5.171 | .612 |
| | TL | SL | 16- 23 | 20.20 | 15.80 | 5 | 1.013 | -4.661 | .631 |
| Poecilia latipinna | SL | TL | 8- 42 | 26.92 | 34.19 | 26 | 1.272 | -.040 | .683 |
| | TL | SL | 10- 54 | 36.29 | 28.32 | 31 | .771 | .328 | .868 |
| **POMACENTRIDAE** | | | | | | | | | |
| Abudefduf saxatilis | SL | | 22- 31 | 26.25 | 32.37 | 8 | 1.068 | 4.327 | .653 |
| | SL | | 22- 31 | 26.25 | 35.75 | 8 | 1.288 | 1.948 | .612 |
| | FL | | 27- 37 | 32.37 | 26.25 | 8 | .907 | -3.103 | .601 |
| | FL | | 27- 37 | 32.37 | 35.75 | 8 | 1.182 | -2.498 | .728 |
| | TL | SL | 30- 42 | 35.75 | 26.25 | 8 | .762 | -.981 | .470 |
| | TL | FL | 30- 42 | 35.75 | 32.37 | 8 | .823 | 2.934 | .608 |
| **POMADASYIDAE** | | | | | | | | | |
| Orthopristis chrysopterus | SL | FL | 11- 46 | 29.42 | 35.10 | 31 | 1.169 | .695 | .970 |
| | SL | TL | 11- 46 | 29.42 | 37.74 | 31 | 1.326 | -1.268 | .518 |
| | FL | SL | 13- 55 | 34.89 | 29.33 | 36 | .852 | -.407 | .737 |
| | FL | TL | 13- 55 | 34.89 | 37.50 | 36 | 1.123 | -1.682 | 1.195 |
| | TL | SL | 14- 59 | 37.77 | 29.49 | 39 | .751 | 1.121 | .425 |
| | TL | FL | 14- 59 | 37.77 | 35.26 | 39 | .892 | 1.587 | 1.017 |
| **POMATOMIDAE** | | | | | | | | | |
| Pomatomus saltatrix | | | 28-108 | 62.82 | 71.26 | 34 | 1.113 | 1.324 | 1.822 |
| | | | 28-108 | 62.82 | 78.53 | 34 | 1.272 | -1.398 | 1.562 |
| | | | 32-106 | 67.52 | 59.55 | 29 | .910 | -1.881 | 1.652 |
| | | | 32-106 | 67.52 | 74.38 | 29 | 1.135 | -2.230 | 2.194 |
| | SL | FL | 35-103 | 68.46 | 54.93 | 28 | .804 | -.119 | 1.067 |
| | SL | FL | 35-103 | 68.46 | 62.61 | 28 | .906 | .583 | .914 |

Table 1.--Relation of standard, fork, and total lengths in marine fishes from coastal Georgia--Continued

| Species | Independent variable X | Dependent variable Y | Size range, independent variable X | $\bar{x}$ | $\bar{y}$ | N | b | a | Sy.x |
|---|---|---|---|---|---|---|---|---|---|
| | | | Mm. | Mm. | Mm. | | | | |
| **SCIAENIDAE** | | | | | | | | | |
| Bairdiella chrysura | SL | TL | 12-151 | 74.48 | 92.97 | 104 | 1.219 | 2.182 | 1.146 |
| | TL | SL | 15-165 | 84.62 | 67.67 | 112 | .815 | -1.270 | .937 |
| Cynoscion nebulosus | SL | TL | 8- 60 | 26.45 | 33.40 | 20 | 1.224 | 1.020 | .867 |
| | TL | SL | 10- 72 | 33.40 | 26.45 | 20 | .815 | -.764 | .707 |
| Cynoscion regalis | SL | TL | 8- 50 | 28.93 | 37.38 | 29 | 1.290 | .070 | .756 |
| | TL | SL | 10- 52 | 33.55 | 25.86 | 29 | .763 | .266 | .714 |
| Larimus fasciatus | SL | TL | 16- 48 | 31.44 | 42.37 | 27 | 1.336 | .362 | 1.079 |
| | TL | SL | 22- 65 | 42.55 | 31.52 | 29 | .751 | -.446 | .702 |
| Leiostomus xanthurus | SL | TL | 12-105 | 48.35 | 61.65 | 71 | 1.288 | -.606 | .910 |
| | TL | SL | 14-111 | 57.70 | 45.24 | 87 | .771 | .760 | .893 |
| Menticirrhus americanus | SL | TL | 7-107 | 54.02 | 68.65 | 94 | 1.252 | 1.028 | .968 |
| | TL | SL | 8-127 | 65.51 | 51.49 | 111 | .797 | -.726 | .827 |
| Menticirrhus littoralis | SL | TL | 7-122 | 59.21 | 73.41 | 102 | 1.216 | 1.393 | 1.062 |
| | TL | SL | 9-148 | 70.03 | 56.32 | 119 | .819 | -1.006 | .781 |
| Menticirrhus saxatilis | SL | TL | 11- 63 | 30.50 | 38.72 | 36 | 1.233 | 1.116 | .679 |
| | TL | SL | 13- 56 | 33.62 | 26.38 | 39 | .814 | -.990 | .537 |
| Micropogon undulatus | SL | TL | 7- 38 | 17.00 | 22.00 | 16 | 1.369 | -1.270 | .810 |
| | TL | SL | 8- 52 | 22.00 | 17.00 | 16 | .728 | .995 | .590 |
| Pogonias cromis | SL | TL | 19- 67 | 33.85 | 43.42 | 7 | 1.316 | -1.133 | .392 |
| | TL | SL | 24- 87 | 43.42 | 33.85 | 7 | .759 | .877 | .297 |
| Sciaenops ocellata | SL | TL | 13- 32 | 21.60 | 27.40 | 5 | 1.247 | .469 | .664 |
| | TL | SL | 16- 40 | 27.40 | 21.60 | 5 | .799 | -.290 | .532 |
| Stellifer lanceolatus | SL | TL | 17- 93 | 54.22 | 71.51 | 27 | 1.295 | 1.295 | 1.298 |
| | TL | SL | 22-123 | 71.51 | 54.22 | 27 | .770 | -.864 | 1.001 |
| **SCOMBRIDAE** | | | | | | | | | |
| Scomberomorus maculatus | | | 15-108 | 57.71 | 64.00 | 35 | 1.102 | .387 | 1.487 |
| | | | 15-108 | 57.71 | 72.23 | 35 | 1.269 | -1.026 | 2.679 |
| | | | 17-103 | 53.59 | 48.47 | 32 | .918 | -.730 | .761 |
| | | | 17-103 | 53.59 | 60.50 | 32 | 1.173 | -2.349 | 2.506 |
| | | | 18-104 | 55.80 | 45.17 | 30 | .812 | -.149 | .643 |
| | SL/FL/TL | FL/TL/SL | 18-104 | 55.80 | 50.07 | 30 | .883 | .771 | .881 |
| **SOLEIDAE** | | | | | | | | | |
| Trinectes maculatus | SL/TL | FL/SL | 9- 50 | 27.82 | 36.42 | 38 | 1.277 | .906 | 1.184 |
| | | | 11- 64 | 34.73 | 26.31 | 48 | .764 | -.209 | .661 |
| **SPARIDAE** | | | | | | | | | |
| Archosargus probatocephalus | | | 11- 15 | 13.00 | 15.20 | 5 | 1.200 | -.400 | .365 |
| | | | 11- 15 | 13.00 | 16.20 | 5 | 1.200 | .600 | .365 |
| | | | 13- 18 | 15.20 | 13.00 | 5 | .811 | .676 | .300 |
| | | | 13- 18 | 15.20 | 16.20 | 5 | 1.000 | 1.000 | .000 |
| | | | 14- 19 | 16.20 | 13.00 | 5 | .811 | -.135 | .300 |
| | SL/FL/TL | FL/TL/SL | 14- 19 | 16.20 | 15.20 | 5 | 1.000 | -1.000 | .000 |

Table 1.--Relation of standard, fork, and total lengths in marine fishes from coastal Georgia--Continued

| Species | Independent variable X | Dependent variable Y | Size range, independent variable X | $\bar{x}$ | $\bar{y}$ | N | b | a | Sy.x |
|---|---|---|---|---|---|---|---|---|---|
| | | | Mm. | Mm. | Mm. | | | | |
| SPARIDAE--Continued | | | | | | | | | |
| Lagodon rhomboides | SL | FL | 14-119 | 66.33 | 77.33 | 15 | 1.138 | 1.873 | .929 |
| | SL | TL | 14-119 | 66.33 | 84.86 | 15 | 1.282 | -.155 | 1.670 |
| | FL | SL | 16-136 | 77.33 | 66.33 | 15 | .878 | -1.597 | .816 |
| | FL | TL | 16-136 | 77.33 | 84.86 | 15 | 1.126 | -2.252 | 1.407 |
| | TL | SL | 18-152 | 84.86 | 66.33 | 15 | .779 | .241 | 1.301 |
| | TL | FL | 18-152 | 84.86 | 77.33 | 15 | .886 | 2.102 | 1.248 |
| STROMATEIDAE | | | | | | | | | |
| Peprilus alepidotus | | | 47-102 | 74.56 | 83.95 | 23 | 1.063 | 4.693 | .777 |
| | | | 47-102 | 73.32 | 101.25 | 28 | 1.450 | -5.057 | 1.320 |
| | | | 54-114 | 83.95 | 74.56 | 23 | .939 | -4.235 | .730 |
| | | | 54-114 | 82.57 | 101.25 | 28 | 1.353 | -10.442 | 1.132 |
| | | | 62-145 | 101.25 | 73.32 | 28 | .687 | 3.751 | .908 |
| | SL | FL | 62-145 | 101.25 | 82.57 | 28 | .737 | 7.918 | .836 |
| SYNGNATHIDAE | | | | | | | | | |
| Syngnathus fuscus | SL | TL | 33-101 | 69.10 | 72.03 | 30 | 1.025 | 1.175 | .404 |
| | TL | SL | 35-105 | 72.03 | 69.10 | 30 | .975 | -1.100 | .394 |
| Syngnathus louisianae | SL | TL | 39- 81 | 55.79 | 58.29 | 24 | 1.039 | .313 | .461 |
| | TL | SL | 41- 84 | 58.29 | 55.79 | 24 | .961 | -.221 | .443 |
| SYNODONTIDAE | | | | | | | | | |
| Synodus foetens | | FL | 32- 44 | 36.07 | 38.30 | 13 | 1.093 | -1.121 | 0.794 |
| | | TL | 32- 44 | 36.07 | 41.23 | 13 | 1.188 | -1.610 | .436 |
| | | SL | 33- 46 | 38.30 | 36.07 | 13 | .883 | 2.263 | .713 |
| | | TL | 33- 46 | 38.30 | 41.23 | 13 | 1.052 | .927 | .873 |
| | | SL | 35- 50 | 41.23 | 36.07 | 13 | .834 | 1.680 | .366 |
| | SL | FL | 35- 50 | 41.23 | 38.30 | 13 | .915 | .575 | .814 |
| TETRAODONTIDAE | | | | | | | | | |
| Sphaeroides maculatus | SL | TL | 9- 37 | 20.17 | 26.71 | 28 | 1.240 | 1.699 | .619 |
| | TL | SL | 12- 47 | 26.71 | 20.17 | 28 | .802 | -1.265 | .498 |
| TRIGLIDAE | | | | | | | | | |
| Prionotus carolinus | SL | TL | 35- 73 | 51.50 | 65.64 | 14 | 1.339 | -3.339 | .820 |
| | TL | SL | 44- 94 | 65.64 | 51.50 | 14 | .744 | 2.690 | .611 |
| Prionotus scitulus | SL | TL | 17- 61 | 37.21 | 46.30 | 13 | 1.248 | -.119 | 1.084 |
| | TL | SL | 22- 78 | 46.30 | 37.21 | 13 | .798 | .258 | .867 |
| Prionotus tribulus | SL | TL | 10- 64 | 39.44 | 50.00 | 9 | 1.264 | .132 | .710 |
| | TL | SL | 13- 81 | 50.00 | 39.44 | 9 | .790 | -.060 | .560 |
| URANOSCOPIDAE | | | | | | | | | |
| Astroscopus y-graecum | SL | TL | 11- 98 | 41.75 | 55.29 | 51 | 1.264 | 2.514 | 1.220 |
| | TL | SL | 16- 96 | 52.86 | 39.64 | 56 | .775 | -1.329 | .946 |

Table 2.--Conversion factors for standard, fork, and total lengths (millimeters) for 82 species of marine fishes occurring in coastal Georgia

[Six factors are given for species with forked caudal fins, two are given for species without forked caudal fins]

| Species | Size range Total length | Standard length to: Fork length | Standard length to: Total length | Fork length to: Standard length | Fork length to: Total length | Total length to: Standard length | Total length to: Fork length |
|---|---|---|---|---|---|---|---|
| | Mm. | | | | | | |
| **ANGUILLIDAE** | | | | | | | |
| Anguilla rostrata | 46- 63 | | 1.676 + 0.988SL | | | -1.498 + 1.008TL | |
| **ARIIDAE** | | | | | | | |
| Galeichthys felis | 51-114 | 1.630 + 1.063SL | -1.245 + 1.320SL | -0.741 + 0.930FL | -2.466 + 1.228FL | -.645 + .781TL | 0.756 + 0.833TL |
| **ATHERINIDAE** | | | | | | | |
| Membras martinica | 16- 93 | .029 + 1.135SL | .154 + 1.199SL | .352 + .874FL | -.033 + 1.061FL | -.078 + .832TL | -.230 + .948TL |
| Menidia menidia | 14-113 | 1.112 + 1.127SL | 1.534 + 1.196SL | -.247 + .875FL | .286 + 1.063FL | -.708 + .827TL | .122 + .934TL |
| **BATRACHOIDIDAE** | | | | | | | |
| Opsanus tau | 71-183 | | .280 + 1.180SL | | | -.051 + .846TL | |
| **BELONIDAE** | | | | | | | |
| Strongylura marina | 23-193 | | 2.878 + 1.061SL | | | -1.676 + .934TL | |
| **BLENNIIDAE** | | | | | | | |
| Chasmodes bosquianus | 14- 84 | | 1.183 + 1.180SL | | | -.666 + .841TL | |
| Hypsoblennius hentzi | 9- 78 | | -.017 + 1.227SL | | | .057 + .814TL | |
| **BOTHIDAE** | | | | | | | |
| Etropus crossotus | 60- 88 | | 1.346 + 1.240SL | | | -.534 + .799TL | |
| Paralichthys dentatus | 32- 69 | | -1.992 + 1.273SL | | | 2.013 + .777TL | |
| Paralichthys lethostigma | 26-117 | | -.005 + 1.257SL | | | -.563 + .800TL | |
| Paralichthys squamilentus | 26- 53 | | 1.501 + 1.223SL | | | -.986 + .812TL | |
| Scophthalmus aquosus | 61- 99 | | 4.225 + 1.212SL | | | -2.502 + .814TL | |
| **CARANGIDAE** | | | | | | | |
| Caranx hippos | 26-132 | .838 + 1.069SL | -.195 + 1.247SL | -.756 + .935FL | -1.246 + 1.166FL | .394 + .800TL | 1.404 + .857TL |
| Chloroscombrus chrysurus | 18-127 | 1.417 + 1.087SL | -1.408 + 1.316SL | -1.352 + .921FL | -3.204 + 1.216FL | 1.412 + .754TL | 2.764 + .822TL |
| Oligoplites saurus | 11-127 | 1.544 + 1.060SL | 1.143 + 1.181SL | -1.324 + .942FL | -.644 + 1.116FL | -.767 + .846TL | .607 + .896TL |
| Selene vomer | 32-105 | -.100 + 1.099SL | -.260 + 1.338SL | .133 + .909FL | -.070 + 1.216FL | .245 + .746TL | .178 + .820TL |
| Trachinotus carolinus | 14-102 | 1.186 + 1.100SL | -1.089 + 1.340SL | -1.461 + .903FL | -3.146 + 1.214FL | 1.064 + .740TL | 2.655 + .822TL |
| Trachinotus falcatus | 16- 79 | 2.386 + 1.082SL | .950 + 1.303SL | -2.148 + .922FL | -2.292 + 1.214FL | -.712 + .769TL | 1.618 + .831TL |
| Trachinotus glaucus | 24- 78 | 1.372 + 1.122SL | -.839 + 1.326SL | -1.171 + .890FL | -2.447 + 1.182FL | .693 + .753TL | 2.106 + .846TL |
| Vomer setapinnis | 53- 81 | -3.061 + 1.169SL | -5.289 + 1.410SL | 2.930 + .850FL | -1.682 + 1.208FL | 4.229 + .702TL | 1.461 + .827TL |
| **CLUPEIDAE** | | | | | | | |
| Alosa aestivalis | 35- 74 | 1.620 + 1.084SL | -1.612 + 1.294SL | -1.445 + .921FL | -3.448 + 1.191FL | 1.292 + .772TL | 3.051 + .836TL |
| Alosa sapidissima | 45- 74 | -.726 + 1.110SL | -3.761 + 1.313SL | .838 + .897FL | -2.738 + 1.179FL | 3.197 + .756TL | 2.758 + .841TL |
| Brevoortia smithi | 27- 95 | 1.992 + 1.050SL | -.191 + 1.285SL | -2.921 + .994FL | -4.188 + 1.263FL | .178 + .776TL | 2.032 + .817TL |
| Brevoortia tyrannus | 18-174 | .098 + 1.103SL | -2.796 + 1.320SL | .389 + .898FL | -2.712 + 1.195FL | 2.531 + .751TL | 2.480 + .834TL |
| Dorosoma cepedianum | 97-150 | 4.605 + 1.056SL | 8.000 + 1.235SL | -3.568 + .939FL | 2.668 + 1.169FL | -5.505 + .802TL | -1.461 + .848TL |
| Dorosoma petenense | 50- 93 | 1.180 + 1.078SL | 2.766 + 1.254SL | -.844 + .923FL | 1.401 + 1.163FL | -1.700 + .790TL | -.927 + .856TL |
| Harengula pensacolae | 56- 79 | 5.083 + .999SL | -.863 + 1.280SL | -4.356 + .988FL | -6.727 + 1.270FL | 1.052 + .776TL | 5.924 + .778TL |
| Opisthonema oglinum | 35- 97 | 1.664 + 1.073SL | .665 + 1.245SL | 1.972 + .864FL | -1.875 + 1.158FL | 4.265 + .735TL | 3.544 + .832TL |
| Sardinella anchovia | 29- 37 | .648 + 1.099SL | -2.417 + 1.304SL | .305 + .881FL | -2.505 + 1.164FL | 2.054 + .761TL | 2.523 + .848TL |
| **CYNOGLOSSIDAE** | | | | | | | |
| Symphurus plagiusa | 13-138 | | -.055 + 1.096SL | | | .096 + .911TL | |
| **CYPRINODONTIDAE** | | | | | | | |
| Cyprinodon variegatus | 21- 51 | | 2.133 + 1.166SL | | | -1.617 + .851TL | |
| Fundulus heteroclitus | 7- 96 | | 1.580 + 1.196SL | | | -.899 + .823TL | |
| Fundulus luciae | 12- 39 | | .378 + 1.210SL | | | -.149 + .820TL | |
| Fundulus majalis | 10-116 | | 1.800 + 1.182SL | | | -1.323 + .844TL | |
| **DIODONTIDAE** | | | | | | | |
| Chilomycterus schoepfi | 15- 64 | | .944 + 1.187SL | | | -.673 + .839TL | |
| **ECHELIDAE** | | | | | | | |
| Myrophis punctatus | 133-172 | | -.923 + 1.013SL | | | .984 + .986TL | |

11

Table 2.--Conversion factors for standard, fork, and total lengths (millimeters) for 82 species of marine fishes occurring in coastal Georgia.-Continued

| Species | Size range Total length | Standard length to: Fork length | Standard length to: Total length | Fork length to: Standard length | Fork length to: Total length | Total length to: Standard length | Total length to: Fork length |
|---|---|---|---|---|---|---|---|
| | Mm. | | | | | | |
| **ELOPIDAE** | | | | | | | |
| Elops saurus | 29-131 | 0.950 + 1.056SL | -2.068 + 1.294SL | -0.844 + 0.946FL | -3.192 + 1.225FL | 1.702 + 0.771TL | 2.730 + 0.815TL |
| **ENGRAULIDAE** | | | | | | | |
| Anchoa hepsetus | 16-117 | .932 + 1.078SL | -.395 + 1.228SL | -.756 + .927FL | -1.508 + 1.141FL | .390 + .817TL | 1.178 + .881TL |
| Anchoa lyolepis | 34- 48 | 1.453 + 1.077SL | -4.971 + 1.378SL | .894 + .866FL | -4.436 + 1.213FL | 4.343 + .707TL | 5.641 + .773TL |
| Anchoa mitchilli | 12- 84 | .389 + 1.078SL | -.315 + 1.214SL | -.350 + .929FL | -.701 + 1.126FL | .768 + .812TL | .586 + .887TL |
| **EPHIPPIDAE** | | | | | | | |
| Chaetodipterus faber | 7- 43 | | 1.000 + 1.278SL | | | -.714 + .779TL | |
| **GERRIDAE** | | | | | | | |
| Diapterus olisthostomus | 23- 72 | .312 + 1.115SL | .398 + 1.318SL | -.226 + .895FL | -.032 + 1.184FL | -.039 + .752TL | .157 + .841TL |
| Eucinostomus gula | 12- 93 | .462 + 1.106SL | -.901 + 1.328SL | .140 + .880FL | -.601 + 1.172FL | .681 + .751TL | 1.304 + .834TL |
| **GOBIESOCIDAE** | | | | | | | |
| Gobiesox strumosus | 3- 68 | | -.027 + 1.240SL | | | .301 + .801TL | |
| **GOBIIDAE** | | | | | | | |
| Gobionellus shufeldti | 17- 67 | | -1.218 + 1.352SL | | | 1.004 + .740TL | |
| Gobiosoma bosci | 9- 56 | | .061 + 1.230SL | | | .246 + .806TL | |
| **MONACANTHIDAE** | | | | | | | |
| Stephanolepis hispidus | 10- 50 | | .568 + 1.247SL | | | -.353 + .795TL | |
| **MUGILIDAE** | | | | | | | |
| Mugil cephalus | 17-219 | 2.1 + 1.135SL | -.424 + 1.266SL | -1.584 + .877FL | -2.733 + 1.115FL | .646 + .787TL | 2.233 + .900TL |
| Mugil curema | 20-155 | 1.448 + 1.157SL | -.468 + 1.288SL | -.542 + .860FL | -1.703 + 1.114FL | .751 + .773TL | 1.488 + .897TL |
| **OPHIDIIDAE** | | | | | | | |
| Rissola marginata | 138-184 | | 8.411 + .973SL | | | -8.231 + 1.026TL | |
| **POECILIIDAE** | | | | | | | |
| Gambusia affinis | 17- 46 | | .796 + 1.276SL | | | .156 + .758TL | |
| Heterandria formosa | 16- 23 | | 5.171 + .951SL | | | -4.661 + 1.013TL | |
| Poecilis latipinna | 10- 54 | | -.040 + 1.272SL | | | .328 + .771TL | |
| **POMACENTRIDAE** | | | | | | | |
| Abudefduf saxatilis | 30- 42 | 4.327 + 1.068SL | 1.948 + 1.288SL | -3.103 + .907FL | -2.498 + 1.182FL | -.981 + .762TL | 2.934 + .823TL |
| **POMADASYIDAE** | | | | | | | |
| Orthopristis chrysopterus | 14- 59 | .695 + 1.169SL | -1.268 + 1.326SL | -.407 + .852FL | -1.682 + 1.123FL | 1.121 + .751TL | 1.587 + .892TL |
| **POMATOMIDAE** | | | | | | | |
| Pomatomus saltatrix | 35-103 | 1.324 + 1.113SL | -1.398 + 1.272SL | -1.881 + .910FL | -2.230 + 1.135FL | -.119 + .804TL | .583 + .906TL |
| **SCIAENIDAE** | | | | | | | |
| Bairdiella chrysura | 15-165 | | 2.182 + 1.219SL | | | -1.270 + .815TL | |
| Cynoscion nebulosus | 10- 72 | | 1.020 + 1.224SL | | | -.764 + .815TL | |
| Cynoscion regalis | 10- 52 | | .070 + 1.290SL | | | .266 + .763TL | |
| Larimus fasciatus | 22- 65 | | .362 + 1.336SL | | | -.446 + .751TL | |
| Leiostomus xanthurus | 14-111 | | -.606 + 1.288SL | | | .760 + .771TL | |
| Menticirrhus americanus | 8-127 | | 1.028 + 1.252SL | | | -.726 + .797TL | |
| Menticirrhus littoralis | 9-148 | | 1.393 + 1.216SL | | | -1.006 + .819TL | |
| Menticirrhus saxatilis | 13- 56 | | 1.116 + 1.233SL | | | -.990 + .814TL | |
| Micropogon undulatus | 8- 52 | | -1.270 + 1.369SL | | | .995 + .728TL | |
| Pogonias cromis | 24- 87 | | -1.133 + 1.316SL | | | .877 + .759TL | |
| Sciaenops ocellata | 16- 40 | | .469 + 1.247SL | | | -.290 + .799TL | |
| Stellifer lanceolatus | 22-123 | | 1.295 + 1.295SL | | | -.864 + .770TL | |

12

Table 2.--Conversion factors for standard, fork, and total lengths (millimeters) for 82 species of marine fishes occurring in coastal Georgia--Continued

| Species | Size range Total length | Standard length to: | | Fork length to: | | Total length to: | |
|---|---|---|---|---|---|---|---|
| | | Fork length | Total length | Standard length | Total length | Standard length | Fork length |
| | Mm. | | | | | | |
| SCOMBRIDAE | | | | | | | |
| Scomberomorus maculatus | 18-104 | 0.387 + 1.102SL | -1.026 + 1.269SL | -0.730 + .918FL | -2.349 + 1.173FL | -0.149 + .812TL | 0.771 + .883TL |
| SOLEIDAE | | | | | | | |
| Trinectes maculatus | 11- 64 | | .906 + 1.277SL | | | -.209 + .764TL | |
| SPARIDAE | | | | | | | |
| Archosargus probatocephalus | 14- 19 | -.400 + 1.200SL | .600 + 1.200SL | .676 + .811FL | 1.000 + 1.000FL | -.135 + .811TL | -1.000 + 1.000TL |
| Lagodon rhomboides | 18-152 | 1.873 + 1.138SL | -.155 + 1.282SL | -1.597 + .878FL | -2.252 + 1.126FL | .241 + .779TL | 2.102 + .886TL |
| STROMATEIDAE | | | | | | | |
| Peprilus alepidotus | 62-145 | 4.693 + 1.063SL | -5.057 + 1.450SL | -4.235 + .939FL | -10.442 + 1.353FL | 3.751 + .687TL | 7.918 + .737TL |
| SYNGNATHIDAE | | | | | | | |
| Syngnathus fuscus | 35-105 | | 1.175 + 1.025SL | | | -1.100 + .975TL | |
| Syngnathus louisianae | 41- 84 | | .313 + 1.039SL | | | -.221 + .961TL | |
| SYNODONTIDAE | | | | | | | |
| Synodus foetens | 35- 50 | -1.121 + 1.093SL | -1.610 + 1.188SL | 2.263 + .883FL | .927 + 1.052FL | 1.680 + .834TL | .573 + .915TL |
| TETRAODONTIDAE | | | | | | | |
| Sphaeroides maculatus | 12- 47 | | 1.699 + 1.240SL | | | -1.265 + .802TL | |
| TRIGLIDAE | | | | | | | |
| Prionotus carolinus | 44- 94 | | -3.339 + 1.339SL | | | 2.690 + .744TL | |
| Prionotus scitulus | 22- 78 | | -.119 + 1.248SL | | | .258 + .798TL | |
| Prionotus tribulus | 13- 81 | | .132 + 1.264SL | | | -.060 + .790TL | |
| URANOSCOPIDAE | | | | | | | |
| Astroscopus y-graecum | 16- 96 | | 2.514 + 1.264SL | | | -1.329 + .775TL | |

13

Table 3.--List of scientific and common names of fishes

[Common name is from American Fisheries Society
Special Publication 2(1960)[1] -a second common
name has been added where the AFS name may con-
fuse species locally]

| Family | Species | Common name |
|--------|---------|-------------|
| ANGUILLIDAE | Anguilla rostrata (LeSueur) | American eel |
| ARIIDAE | Galeichthys felis (Linnaeus) | Sea catfish |
| ATHERINIDAE | Membras martinica (Valenciennes) | Rough silverside |
| | Menidia menidia (Linnaeus) | Atlantic silverside |
| BATRACHOIDIDAE | Opsanus tau (Linnaeus) | Oyster toadfish |
| BELONIDAE | Strongylura marina (Walbaum) | Atlantic needlefish |
| BLENNIIDAE | Chasmodes bosquianus (Lacépède) | Striped blenny |
| | Hypsoblennius hentzi (LeSueur) | Feather blenny |
| BOTHIDAE | Etropus crossotus Jordan and Gilbert | Fringed flounder |
| | Paralichthys dentatus (Linnaeus) | Summer flounder |
| | Paralichthys lethostigma Jordan and Gilbert | Southern flounder |
| | Paralichthys squamilentus Jordan and Gilbert | Broad flounder |
| | Scophthalmus aquosus (Mitchill) | Windowpane |
| CARANGIDAE | Caranx hippos (Linnaeus) | Crevalle jack |
| | Chloroscombrus chrysurus (Linnaeus) | Bumper |
| | Oligoplites saurus (Bloch and Schneider) | Leatherjacket |
| | Selene vomer (Linnaeus) | Lookdown |
| | Trachinotus carolinus (Linnaeus) | Pompano |
| | Trachinotus falcatus (Linnaeus) | Permit |
| | Trachinotus glaucus (Bloch) | Palometa |
| | Vomer setapinnis (Mitchill) | Atlantic moonfish |
| CLUPEIDAE | Alosa aestivalis (Mitchill) | Blue back herring |
| | Alosa sapidissima (Wilson) | American shad |
| | Brevoortia smithi Hildebrand | Yellowfin shad |
| | Brevoortia tyrannus (Latrobe) | Atlantic menhaden |
| | Dorosoma cepedianum (LeSueur) | Gizzard shad |
| | Dorosoma petenense (Günther) | Threadfin shad |
| | Harengula pensacolae Goode and Bean | Scaled sardine |
| | Opisthonema oglinum (LeSueur) | Atlantic thread herring |
| | Sardinella anchovia Valenciennes | Spanish sardine |

---

[1] A list of common and scientific names of fishes from the United States and Canada,
by Reeve M. Bailey, et al. 1960. Amer. Fish. Soc. Spec. Publ. No. 2 (2nd ed.), 102 p.

14

Table 3.--List of scientific and common names of fishes--Continued

| | | |
|---|---|---|
| CYNOGLOSSIDAE | Symphurus plagiusa (Linnaeus) | Blackcheek tonguefish |
| CYPRINODONTIDAE | Cyprinodon variegatus Lacépéde | Sheepshead minnow |
| | Fundulus heteroclitus (Linnaeus) | Mummichog |
| | Fundulus luciae (Baird) | Spotfin killifish |
| | Fundulus majalis (Walbaum) | Striped killifish |
| DIODONTIDAE | Chilomycterus schoepfi (Walbaum) | Striped burrfish |
| ECHELIDAE | Myrophis punctatus Lutken | Speckled worm eel |
| ELOPIDAE | Elops saurus Linnaeus | Ladyfish |
| ENGRAULIDAE | Anchoa hepsetus (Linnaeus) | Striped anchovy |
| | Anchoa lyolepis (Evermann and Marsh) | Dusky anchovy |
| | Anchoa mitchilli (Valenciennes) | Bay anchovy |
| EPHIPPIDAE | Chaetodipterus faber (Broussonet) | Atlantic spadefish |
| GERRIDAE | Diapterus olisthostomus (Goode and Bean) | Irish pompano; mojarra |
| | Eucinostomus gula (Quoy and Gaimard) | Silver jenny; mojarra |
| GOBIESOCIDAE | Gobiesox strumosus Cope | Skilletfish |
| GOBIIDAE | Gobionellus shufeldti (Jordan and Evermann) | Freshwater goby |
| | Gobiosoma bosci (Lacepede) | Naked goby |
| MONACANTHIDAE | Stephanolepis hispidus (Linnaeus) | Planehead filefish |
| MUGILIDAE | Mugil cephalus Linnaeus | Striped mullet |
| | Mugil curema Valenciennes | White mullet |
| OPHIDIIDAE | Rissola marginata (DeKay) | Striped cusk-eel |
| POECILIIDAE | Gambusia affinis (Baird and Girard) | Mosquitofish |
| | Heterandria formosa Agassiz | Least killifish |
| | Poecilia latipinna (LeSueur) | Sailfin molly |
| POMACENTRIDAE | Abudefduf saxatilis (Linnaeus) | Sergeant major |
| POMADASYIDAE | Orthopristis chrysopterus (Linnaeus) | Pigfish |
| POMATOMIDAE | Pomatomus saltatrix (Linnaeus) | Bluefish |

Table 3.--List of scientific and common names of fishes--Continued

| Family | Species | Common name |
|--------|---------|-------------|
| SCIAENIDAE | Bairdiella chrysura (Lacépède) | Silver perch; yellowtail |
| | Cynoscion nebulosus (Cuvier) | Spotted seatrout |
| | Cynoscion regalis (Bloch and Schneider) | Weakfish; gray seatrout |
| | Larimus fasciatus Holbrook | Banded drum |
| | Leiostomus xanthurus Lacepède | Spot |
| | Menticirrhus americanus (Linnaeus) | Southern kingfish |
| | Menticirrhus littoralis (Holbrook) | Gulf kingfish |
| | Menticirrhus saxatilis (Bloch and Schneider) | Northern kingfish |
| | Micropogon undulatus (Linnaeus) | Atlantic croaker |
| | Pogonias cromis (Linnaeus) | Black drum |
| | Sciaenops ocellata (Linnaeus) | Red drum; channel bass |
| | Stellifer lanceolatus (Holbrook) | Star drum |
| SCOMBRIDAE | Scomberomorus maculatus (Mitchill) | Spanish mackerel |
| SOLEIDAE | Trinectes maculatus (Bloch and Schneider) | Hogchoker |
| SPARIDAE | Archosargus probatocephalus (Walbaum) | Sheepshead |
| | Lagodon rhomboides (Linnaeus) | Pinfish |
| STROMATEIDAE | Peprilus alepidotus (Linnaeus) | Southern harvestfish |
| SYNGNATHIDAE | Syngnathus fuscus Storer | Northern pipefish |
| | Syngnathus louisianae Günther | Chain pipefish |
| SYNODONTIDAE | Synodus foetens (Linnaeus) | Inshore lizardfish |
| TETRAODONTIDAE | Sphaeroides maculatus (Bloch and Schneider) | Northern puffer |
| TRIGLIDAE | Prionotus carolinus (Linnaeus) | Northern searobin |
| | Prionotus scitulus Jordan and Gilbert | Leopard searobin |
| | Prionotus tribulus Cuvier | Bighead searobin |
| URANOSCOPIDAE | Astroscopus y-graecum (Cuvier) | Southern stargazer |

MS. #1756
GPO 860-570

16

CPSIA information can be obtained
at www.ICGtesting.com
Printed in the USA
BVHW041131310119
539142BV00014B/1306/P